To: Claudia
From: Don Bell
May 29, 2021

BARGE NOTES

The Journal of a Yukon River Deckhand

By Don Bell

BARGE NOTES

All rights reserved

Copyright ©2016 by Don Bell

ISBN-13: 978-1533136602
ISBN-10: 1533136602

DEDICATION

For my mom and dad and barge men everywhere.

Also dedicated to Yutana* Barge Lines. They always got the job done.

"Barging ahead since 1916."

*Yutana is derived from two words – Yukon and Tanana

There's a race of men that don't fit in,

A race that can't stay still;

So they break the hearts of kith and kin,

And they roam the world at will...

- Robert William Service
  (THE MEN THAT DON'T FIT IN)

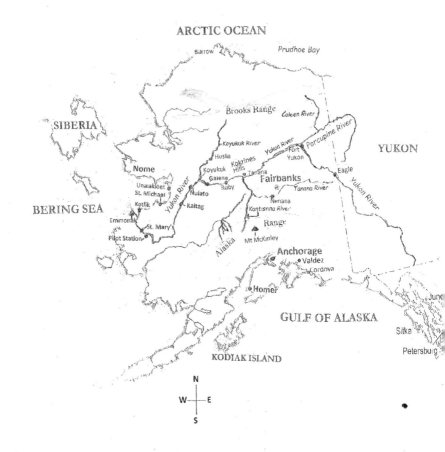

CREW

MV Pat

Captain Murphy
Pilot Wayne
Engineer Allen
First Mate Virgil
Deckhand Don
Deckhand Dave
Cook Kathy
Shop Foreman Joseph

MV Rampart

Captain McCarthy
Pilot Casey
Engineer Kevin
First Mate Curtis
Deckhand Don
Deckhand Nate
Cook Hazel

INTRODUCTION

Barge Notes is a true story. It took place in Alaska over thirty years ago while I worked as a deckhand on the Yukon River. I wrote down what happened every day with the date on whatever was available at the time – pieces of paper or cardboard I found on deck, sometimes brief reminders on the palm of my hand when paper was not available. I wrote some of it in a notebook while off duty in my bunk, and I kept everything in my duffle bag. During the winter I transferred this journal in complete sentences to several permanent notebooks.

This book is the record of that time as it happened. I changed only the names of crew members to protect privacy along with a few transitional timelines.

- Don Bell, May 2016

BARGE NOTES

On Thursday night I got drunk in the Wonder Bar and fell in the Chena River. It is springtime thaw in Fairbanks known as breakup and after leaving the bar I shared some whiskey in the willow brush across the street with a native girl and fell in the river when I slipped in the mud. I climbed out and wandered through town for a while trying to find my hotel. I flew from Fairbanks to Nome early Saturday morning and was tired because the drunks in the parking lot of the hotel were fighting all night. I observed they have a very limited vocabulary.

May 1 – I took Munz Airlines from Nome to St. Michael today. This Eskimo village is Yutana Barge Line's lower Yukon River port and shipyard. It is a collection of one-level wood frame houses set out on the Arctic coastal plain. The ground is covered with snow and the pack ice hugs the shore.

May 2 to May 7 – On Sunday I began my first day as a deckhand on a tugboat named the *MV Pat.* The MV stands for motor vessel. When the remainder of the crew showed up I spent most of this day resting and visiting with them. I worked all week scraping rust from the air voids and painting this old tugboat that is sitting on greased timbers in the mud a few feet from the frozen Bering Sea. We are all working to get her ready so she can do the job during the next navigation season. At this point it looks to me like she will not float.

Snowmelt water is flowing down the muddy streets of the village under and around the machine shop and cookhouse. There are no cars here and just a few pickup trucks that look like they have had some pretty rough service.

The sky is gray one day, then cold and sunny, snowing the next day and then a cold fog moves inshore off the pack ice. After work I've been taking long walks out to the airfield to collect my thoughts and prepare my mind for the next work day on this lonely Arctic coast. Aircraft are not able to land here now as the airfield has thawed out and become a mud hole. Thoughts of my girlfriend in Portland, Oregon bother me because she says I am too restless for a stable relationship and she does not want to see me anymore.

May 8 – It is Saturday night and we get the day off tomorrow because it is Mother's Day. We usually work seven days a week. There is a mix of wet snow and icy rain moving in on us from the northeast and our rooms on the tugboat are wet. Water seeps in under the doors and through cracks in the windows. The mattresses are wet and there is water on the floors. This boat looked like a wreck when I first laid eyes on her and so far she is functioning just the way she looks.

I shared some whiskey with Captain Murphy, Engineer Allen and Kathy, the cook, tonight and we were able to laugh at our situation. Captain Murphy is a swarthy person with a big black beard and he looks like he should be living in a cave. Pilot Wayne, First Mate Virgil and the other deckhand, Dave, are cleaning their rooms. The Lord only knows what fate is waiting for us when we put this old tugboat in the water.

May 9 – It is Mother's Day on the Arctic Coast with a cold, wet east wind. I wish I could call my mom but it is not possible here. We have been working on the *Pat* for ten days now and she is beginning to look a little better on the inside.

May 10 to 12 – On this Monday night there is much sadness in the village as our shop foreman's brother was found dead in a pool of snowmelt water in the frozen tundra several hundred yards from his cabin. He passed out drunk in the early morning and died of exposure. This tragedy has made the moaning wind seem even more cold and lonely. Later on I walked through the village on the boardwalk over the mud with Kathy this melancholy evening.

The melting snow is revealing trash in front of the cabins that was discarded during the winter. When we returned to the boat we met other crew members standing in different places, studying our tugboat home which is the most important thing in our lives right now. I am developing an affection for the *Pat*.

Every evening after dinner I walk out into the tundra where I can watch and listen to the sea birds returning to the

Arctic Coast for the summer. Geese, snipes, widgeons and curlews call out constantly to form their bonds for the breeding season.

Eskimo women wearing long dresses called *kuspuks* are fishing for tomcod through cracks in the sea ice. They use safety pins for hooks and bright thread for bait as the little cod will bite on anything. These fish are very tasty after frying whole and eaten like potato chips.

I got mad at Pilot Wayne yesterday when he told me to hurry my breakfast and fuel the forklift. I was sorry about it later. I woke up in the morning with a bad attitude as in my dreams I was lost and I couldn't find my way home. It is a recurrent dream and I think it is because I don't know where I belong. Maybe this job will help me to find out.

Today I scraped scale and rust from inside the boat's air voids and pumped out the bilge water. It is a dirty job but someone has to do it and that someone is the deckhand.

May 13 – I have been operating a needle gun which is powered by compressed air and uses the action of metal pins to pound and remove rust from the barge's hull. It is the loudest machine I have ever encountered and I believe, if it is possible to get an ear infection from noise, this machine will do it. We'll begin painting tomorrow and I am starting to think we can get the *Pat* ready for another navigation season. We also had a safety meeting today but it didn't do any good because when it was over engineer Allen used a cutting torch on a metal table next to the fuel tanks. I stood by with a fire extinguisher in case the boat blew up but it would be too late at that point.

I took a walk through the village after work like I do every night. The Eskimo kids have wonderful expressive round faces and they always say hello to me. Two little girls were floating around in a mud hole in front of the school on a boat made from a piece of scrap Styrofoam. Junk and garbage is everywhere. Disposable diapers are blown out into the tundra by the wind and get caught up in the willow brush. One of the Eskimos got drunk and shot at them because the diapers are white in color like the ptarmigan which are the wild grouse of the region. The drunk got frustrated when the "ptarmigan" refused to fall out of the bushes.

Guns and alcohol are a bad combination. St. Michael is a dry village and no products of any form containing alcohol are sold in the native store but some people hide it in their luggage when they return from a visit to Nome. A bottle of illegal rum sells for one hundred fifty dollars in the village.

May 14 to 19 – We've been painting the barge decks red and the boat voids yellow this week. I've been cleaning the inside of the fuel barge and using chromate paint in the voids. The fumes are very unpleasant and I am completely covered with a gas mask and duct tape. It is a nasty job and I will be glad to finish it.

After lunch on Thursday I fell into an open hatch that put a bruise the size of a grapefruit on my thigh. It is very sore and I am lucky I did not break my leg. The days are clear with bright sunlight and a cold north wind that quits in the evening. The curlews in the tundra at North Beach call out to each other constantly. There is more standing water on the pack ice every day and we received word on the marine radio that the ice slid out of the Tanana River four days ago and the Yukon River ice has broken up all the way to Stevens Village. The barges in our upper river port of

Nenana will be in the water soon. The sunset tonight was a soft orange glow as seen through a cold mist due to the sun's reflection off the pack ice.

May 18 – Today I stepped on a bare spot in the power cable that runs from the machine shop to the tugboat. The cable was smoking in the mud and the jolt lifted me straight up about three feet. The problem was caused by Captain Murphy who has been driving across the power line repeatedly with a front-end bucket loader. I am beginning to wonder about his judgment. Joseph, the shop foreman, is wrapping new insulation around the wire before someone is killed.

There are two bullet holes in the floor of my room on the boat where the former engineer, "Dirty Neck Pete," got drunk and tried to shoot Captain Murphy in the foot last year. I am beginning to understand why someone might want to do that.

We've been watching news at night on satellite television. Everything seems very far away and unimportant out here.

I sure would like to get some mail but I need to forget about my girlfriend in Oregon and the mail as neither one is reliable.

May 19 to 25 – On Friday I went out on the pack ice near Whale Island to fish for tomcod. I caught twelve of the little fish but I did not pay attention to the weather and when I looked up I was surrounded by a thick wall of freezing fog that had crept up on me and erased my sense of direction. Sea water from a rising tide was surging up through the cracks and spreading out over the ice and, when I realized I could not remember how to get back to land, I felt the chill of fear in my chest. Within a few minutes I heard voices and two Eskimos emerged from the fog dragging a sled with a piece of rope. The older Eskimo was alarmed to see me and he said, "It's time to get off the ice. The tide is coming in and I get scared out here when that happens now that I am older. It used to be fun to stay longer but I know better now. If you get lost on the ice you will die."

I was sure glad to see those guys and I followed them back to shore and vowed never to get caught out there like that again.

On Saturday we took the day off as we are close to being finished with our preseason maintenance work on the barges. I had three tomcod for breakfast.

We will be back to work on Sunday. All of us needed a day off.

May 25 – On this evening I walked out to the north beach and was glad to see a long lead of open water as I know it won't be long before we sail away from here and I am ready.

After work the following day our welders cut two animal silhouettes from a piece of sheet metal. They set up a turkey and an armadillo in the gravel as targets for a Thompson .45-caliber submachine gun. This gun was a weapon of choice for Al Capone and the Chicago mobsters during the Prohibition era and we had fun firing at the metal animals until the local Eskimo sheriff arrived and told us to stop because the noise was alarming the people in the village. These barge workers are a wild bunch and I am feeling pretty wild myself.

During the next two days I repaired and greased the tugboat's winches and put 300 feet of 5/8-inch wire cable on the center winch which will also be used for the anchor.

May 26 – The wind has blown hard out of the southwest for three days and the pack ice is melting fast. Herring, smelt and tomcod are crowding into the shallows by the millions. On this day it rained and I learned how to splice cable with a marlin spike.

May 27 – When I woke up this morning all the pack ice had disappeared in the night. St. Michael Bay was completely clear and when I walked out to the north beach I could barely make out the jagged edge of the pack ice far out to sea across the open water of Norton Sound. When I suggested to Joseph, the Eskimo shop foreman, that we should put the boat in the water and get out of here, he said the ice would come back. I did not believe him until the next morning when I was amazed to see all the pack ice broken into large plates and shoved back tight against the shore again. Joseph said when the ice goes out again it will not come back and after two days the pack ice disappeared overnight and I never saw it again.

May 29 to June 6 – Today Joseph pushed the *Pat* into St. Michael Bay with a dozer but the rudder was jammed and Captain Murphy could only make left turns. He drove the boat around in circles until I was directed by him to put on a survival suit and force my way under the stern until I could feel the rudder and find out why it would not function. After several attempts holding my breath and groping in the freezing sea I determined that a large timber was lodged against the rudder. When I got back in the boat I was shaking from the cold and I don't think even with a survival suit that anyone could live very long in this ocean.

    We attached a line to the *Pat* and after hauling her out of the water the timber was released but upon inspection it was discovered the *Pat* was leaking and the welders went to work patching holes in the hull again. I don't know how long this boat will float as we may sink.

    After this repair Joseph pushed the *Pat* back into the bay with the dozer and the next day we filled our boat fuel tanks at the Chevron tank farm. The fueling dock looks like it is in a state of imminent collapse as does just about everything else out here. We took on fresh water in the afternoon from a nearby creek with the fuel hose as directed by Captain Murphy. This water leaves an unpleasant oily film on the skin after taking a shower and makes the worst coffee I've ever tasted.

June 7 – We are heading out to sea today and down the shore through Stephens Pass on our way to the mouth of the Yukon River and the village of Kotlik. After every task being as hard as possible we are finally underway. We have just passed several commercial herring boats and a big processor ship in the twilight of midnight. There is something about the lights on these boats that makes me lonesome.

June 10 – We are on our way back to St. Michael from Kotlik with two ocean barges named *OB-5* and *The Butte* and, after some trouble finding the channel at the north mouth of the Yukon River, Wayne marked it from the pilot boat with an orange buoy.

Upon entering Norton Sound we observed that the ocean looked a little rough past Point Romanoff, and Captain Murphy decided to switch from pushing our barges to towing them as the ocean swells would break our tackle if we did not do this and the barges would no longer be under control. It is important that this procedure is accomplished in calm water as the deckhand must transfer himself to the barges in order to attach the towline and return to the tugboat when it is brought up alongside the barge. I could see that attempting this operation in rough seas would be difficult and dangerous because the barge and tugboat would be heaving in the swells and the deckhand must jump from the tugboat to the barge and back again exactly when they come together to avoid falling into the ocean and getting crushed between the two vessels. I had no trouble on this occasion but I felt some concern while riding the swells on the barges alone when they were no longer under control of the tugboat. We began towing our two barges in tandem with hawser lines at 3 p.m. and I was relieved that I had accomplished my first ocean towing mission without an accident.

  When I returned to the tugboat I was informed that the toilet was plugged and was directed by the First Mate Virgil to shoot a fire extinguisher into the bowl as this would have the effect of removing the obstruction. This procedure solved the problem and when I was relieved of my duties I watched a movie called "Used Cars" on our video cassette machine in the galley with Joseph. There are no cars in St. Michael and Joseph has never even had a driver's license. The Eskimos tend to find antics of the white people who are rushing around in the city very comical and way out here, detached as I am from the rest of the world, I can see absurdities in modern life I had not noticed before. Life is much less complex here. The other world of cars, traffic jams, asphalt and the frenzy of movement in the city seems very far away and without much meaning to me in this place.

Even though the abuse of alcohol and the introduction of television has compromised this culture and their simple life in intimate association with the sea, the Eskimos are resilient people and the weather conditions and rhythms of nature still dominate their activities.

After watching our movie I washed my face to the clacking sound of the compressed air pump and went to my little room behind the pilot house. The throbbing of the engines is a comfort to me as we pass up the coast to Biscuit Island on our way to St. Michael. The setting sun was bright red at midnight as we moved past the big fish buyer ship with its glaring white quartz light on the mast.

June 11 – It was a clear morning today with a steady north wind of 25-30 knots. We anchored *The Butte* barge and *OB-5* out in the bay with pennant lines while Pilot Wayne moved the *Iron Monster* barge over to the fuel farm which is the facility for the onshore bulk fuel tanks. We will use the *Iron Monster* as a refueling station. It is sometimes amusing to watch Captain Murphy trying to explain something to the manager of the fuel farm by yelling over the noise of the transfer pump. Because the command to "turn it off" is commonly followed by the response "What did you say?" Constant visual contact and the use of hand signals is a more effective means of communication.

A herring boat named the *Ms. Behavin* was tied up to the *Iron Monster* barge when a Southeaster roared through this afternoon and whipped the calm green water of the bay into an angry sea of heaving gray-colored swells. Dark slanted clouds brought a heavy rain and the wind banged the *Iron Monster* and the fishing boat together until the weather cleared to provide a pleasant evening that held a strong clean smell of the ocean.

At midnight we loaded 240,000 gallons of assorted fuels and departed across Norton Sound for the town of Unalakleet. When we approached the mouth of the Unalakleet River an old Eskimo pilot came aboard to guide us into port but our barge is heavy and we had to wait for the rising tide. We grounded on a sandbar and Captain Murphy decided to drop the anchor while we waited, against the advice of the Eskimo pilot who said it would be very difficult to retrieve from the mud. The pilot was correct as the winch did not have the power to raise the anchor and Captain Murphy tried to jerk it free with the tugboat. This procedure came very close to breaking the anchor cable which could have killed or injured one of us. The eye of the cable was starting to come apart when the anchor came up. I am not afraid of dangerous work but I don't want anyone to get killed because of incompetence. I spoke to Pilot Wayne and engineer Allen and they agree with me but they don't want to say anything and I will try to control my temper as I am beginning to think Captain Murphy has some of the qualities of an idiot.

June 12 – On this morning we pushed our fuel barge into the Unalakleet River following the wake of a herring boat named the *Donna Marie*. It took all day and into the night to pump our variety of fuels into the onshore tanks. The weather is cool and there is a group of herring boats in Norton Sound with their nets strung out in the rain.

Since there is a lot of money to be made selling herring eggs, there are fishermen here from many other places. Most of them are young men with beards and there are some women but because of their work clothes and hats the only way I can tell if someone is a woman is when they don't have a beard.

I went into the Alaska Commercial Company store to buy a new starter rope for the 40-horse Mercury Pilot boat motor and a knife to replace the one Deckhand Dave accidentally dropped overboard yesterday. It did me good to get away from the tugboat for a while and look at some new faces.

June 13 – We finished pumping fuel yesterday and took the night off in Unalakleet. The cool, quiet evening on the river with the soft orange glows of the midnight sunset was very pleasant. In the morning we spun out to sea in the channel between the bluff and the herring nets on our way back to St. Michael. The sea looked a little rough at first but we kept pushing our barge through the swells and the trip went well and the crew was relaxed. We passed fishing boats named *Invader*, *Impala*, *Viking Queen*, and *Mitrofania*.

    I was in the pilot house visiting with Wayne and Dave when a herd of walrus surfaced about thirty yards off the starboard bow. Their huge heads appeared suddenly with long white tusks flashing in the sun while they took a look around and disappeared with flippers splashing into the clear, green Bering Sea. We were just off Egg Island when a whale surfaced off the port side blowing spray into the air, and with a rising tail he slipped back beneath the waves.

    June 14 to 18 – Our next trip to Unalakleet was easy with calm seas but on our way back to St. Michael the wind increased and due to heaving swells we had to begin towing. Captain Murphy almost waited too late to do it safely and on this occasion I heard him say he does not like the sea and I believe it frightens him. I watched another herd of walrus when we passed Egg Island again and I must say these creatures are very inquisitive and it is fascinating to observe an animal I have never seen before.

Upon our return to St. Michael we made preparations for our trip to the north mouth of the Yukon River and the town of Kotlik. We loaded deck freight, numerous barrels of oil, Arctic engine gear lube, aviation gasoline and several pallets of Blazo fuel and starting fluid. We also loaded several small boats, engine parts, and 300,000 gallons of diesel fuel.

After pushing our barges out of town today on the 18th of June, we ran into heavy swells and had to switch to towing. I was off duty and as I just wanted to check out for a while I smoked some weed and took a few nips on a bottle of whisky in my room. I felt the boat getting shoved around and there came a frantic knock on my door. It was Dave and he said, "We're going to tow. We need your help."

It was bad timing for me as I needed all my physical and mental abilities for the towing operation and these qualities had been compromised at that point. I damn near fell in the ocean when jumping back and forth from the tugboat to the barges for the purpose of attaching the towing lines and it was a very unpleasant feeling to be standing on a heaving barge while under the influence of marijuana.

Kathy and Dave were seasick and Wayne took over driving the boat after Captain Murphy left the pilot house in distress saying the sea had made an old man out of him. I stood by the rail outside the pilot house riding the swells and feeling the spray in my face and watching Stephens Pass approach and St. Michael fade in the distance.

June 19 – After moving through Stephens Pass between Stewart Island and the mainland, the ocean calmed and red streaks of light from the setting sun were reflected in the soft velvet surface of the sea. After midnight we went back to pushing our barges and entered the Kotlik River about 8 a.m. where we tackled up to another tugboat named *MV Independence*. She has been rebuilt with a wider deck and long narrow structure with very little living space and a lot of money spent on her rehabilitation. We transferred 40,000 gallons of gasoline to the Kotlik

Corporation's shore facility after a great deal of effort dragging the fuel hose through the mud.

June 20 – First Mate Virgil woke up Captain Murphy this morning by operating a needle gun against the door to his room. That would be a very unpleasant way to wake up as the noise is extremely intense and irritating. Captain Murphy returned the favor by throwing plastic water bottles from the pilot house down into the galley on Virgil while he was having breakfast.

At midnight we pushed through Apoon Pass and arrived in the town of Emmonak at 9 a.m.

An Eskimo showed up today with his wife and eight kids in an old wooden boat to visit with our engineer. I didn't think ten people could safely fit into a boat that size but the Eskimos are experts with boats and their capacity. The shoreline here is covered with thick alder and willow brush and I stepped out on deck this evening to watch the brilliant orange glow of the sunset shining on the churning river with our tugboat leaving a frothy trail behind us.

June 21 – We pushed through Kwikpak Pass this morning and on into the

Head of Passes near the south mouth of the Yukon River where we ran aground and almost broke tackle with the *OB-5* barge on the starboard side. The barge was beginning to get away from us when I caught her with a springline and brought her back. We are moving slowly now as the channel often changes here. A Japanese fish processor ship named *Makuhanna Maru* was anchored off our port side when we passed.

June 22 – We pulled into the town of Alakanuk after pumping fuel into the Emmonak Corporation's shore tanks yesterday. It was a hassle getting the unloading timbers into position as we did not have enough chain so I just rammed them into the mud with the forklift. These timbers provide a ramp to span the distance between the barge and shore. My patience is wearing thin with the lack of necessary equipment and I hope I can contain my temper.

In the afternoon I transferred diesel fuel to *Bering Sea Processors* and *Schencks Seafood Corporation.* They are two factory ships that are taking on salmon and, after securing our barge to the ship, a crew of young guys showed up to help us pull the hose. They were very enthusiastic and I believe they enjoyed the action and new faces. These guys came north from all over the lower forty-eight states for the money to be made in the sale of Yukon River salmon and it was good to visit with them.

June 23 – We are now into the main channel of the Yukon River and on our way to Mountain Village. There are numerous small boats scattered about in the river with their crews pulling gill nets in the rain under a dark summer solstice sky. Salmon season began this night at 6 p.m.

June 24 – We arrived in Mountain Village today and the US Public Health Service did not want the big tracked excavator unloaded yet so we transferred fuel to the shore tanks during the night. This country is openly divided with patches of alder brush, black spruce and scattered small birch trees. It reminded me of central Montana until I looked more closely at the village with the cemetery on the hill, the dusky children and the many small boats tied up along the shore. Incongruous federal government houses and sewer system projects stand in contrast to the simplicity of the old log cabins.

We entered the Andreafsky River about noon today, took on fresh water, washed the decks, pulled into the nice new docks at the town of St. Mary's at 1 p.m. and transferred fuel into a large blue shore tank for Trinity Seafood Corporation. An Eskimo employee sat on top of the tank and peered into the opening in an effort to determine how much fuel was already in the tank. He did not have a sounding stick and, after pumping a thousand gallons, a geyser of fuel blew out the top in the Eskimo's face and knocked him off the tank. I shut down the pump quickly but not before a torrent of diesel fuel ran down the hill and into the river and while the Eskimo was trying to stand up in the fuel soaked mud. Captain Murphy dumped all of our laundry detergent into the river in an attempt to soak up the spill. It was a hell of a mess and as there was more fuel in the tank than was reported to us, this was just one more problem.

A large fish-hauling boat passed us with a gigantic sign on board that said "CASH." Everyone is trying to outbid everyone else for the purchase of salmon. St. Mary's is a pleasant town with an old Catholic mission house which has a peaceful atmosphere in contrast to the condition of this deckhand who is developing a bad attitude.

After the fueling debacle, Dave asked me where the pig was and I told him I didn't know. This was not a good sign. The pig is a foam plug that is used to clean fuel from the hose after the transfer is complete. The pig is shoved through the fuel hose by compressed air and must be removed before the next fueling operation or it becomes jammed into the end of a hose full of fuel which is extremely heavy and difficult to handle. During this conversation Dave was waving his arms wildly as a cloud of mosquitoes had attacked him in great numbers. I said, "Are the skeeters bothering you?" He said, "No, they are just flying around," and I realized further comment on this subject was useless.

Captain Murphy had started the fuel transfer pump without my knowledge before the pig was removed. The hose was full of fuel and we had to move it back to the barge in that condition. It was extremely heavy and the chore was very difficult. I confronted Captain Murphy and, when he said this problem was all part of the job, I grabbed him and threatened to throw him into the river.

Dave ran up to me and said, "Don't do it. He'll drown" and I let him go. Captain Murphy screws up too much and makes the work a lot harder than it should be. I feel like I have to throw him in the river or walk off the boat before I get into big trouble with my temper.

June 24 – We arrived in the village of Pilot Station today and after filling the village tanks with heating oil I told Wayne and Dave I am leaving the boat because when I see Captain Murphy again I don't know what I will do. They said they will be sorry to see me go but they understand the danger.

I left the boat at noon in Pilot Station and watched the *Pat* push the barge upriver without me. A native guy who had been watching us transfer fuel came up and asked me if I had missed the boat. I explained that I left the boat because of a disagreement with the Captain and on the suggestion of the native guy I had a job on a pipe-laying crew for the US Public Health Service within forty-five minutes in Pilot Station.

We are putting in a sewer line from a new federal government housing project on the hill to a waste treatment settling pond below the road.

Pilot Station is a small quiet village where everyone goes about their life on the river catching and smoking salmon. The people come over to visit and the kids play outside even when the mosquitoes are the thickest.

June 25 to July 10 – This job is slow and relaxed and I have been assigned living quarters in an old house with two other guys on the crew.

Joe is a quiet white guy from Idaho and Basil is a talkative black guy from Yakima, Washington. I told him I'd never known anyone named Basil and he said he had never known anyone named Basil either. He said he grew up in Chicago and his mother liked to cook with spices. He wanted to be a cowboy so he went to Montana and got into trouble with some white guys who talked him into helping them steal some cows. They got caught and he ended up in prison for 18 months. "You don't want to be a black man in a prison in Montana," he said. He warned me about getting mixed up with white guys. "They'll get you into trouble," he said and winked at me.

Yesterday the boss told Basil to take the three-wheeler down to the shop and get some tools. I was in the ditch with a cloud of mosquitoes laying pipe when Basil drove by too fast on loose gravel. The three-wheeler is a dangerous vehicle on any surface and Basil lost control and flew off the road. The three-wheeler with Basil rolled over into the treatment pond and I laughed so hard I could barely breathe. He was wet but unhurt except for a few scratches and was grateful the pond was not yet in use and was full of clean rainwater instead of the contents of toilets.

During the Fourth of July there were no barbecues or fireworks displays. The native people went about the business of drying and smoking salmon as usual and because we did not work that day it was nice to visit with some of them.

July 15 – On Thursday the boss sent me on the mail plane to the town of Bethel at the mouth of the Kuskokwim River to receive a special order of pipe joints and valves from a Crowley ocean barge. Bethel is a regional center for sea and air transportation and I had a fine time driving around in an old U.S. Government pickup truck, eating out in restaurants and looking at women. It felt good to be on my own again. I cashed my paycheck at a bank and returned to Pilot Station in the morning on Sunday. I walked back up to the airfield in the afternoon where I looked out over the hills, forest and river and felt some unexpected remorse for leaving the *Pat* and her crew.

The next day a new boss named Raymond showed up and yelled at me after he ordered me to get a wrench and I had stopped first for a few minutes to help Basil pick up a heavy sewer pipe. I told this new boss not to talk to me that way and he backed off. I want to control my temper but if he does that again there will be trouble. I spoke to Basil and Joe about this incident after work and they said just do what he says and don't let him get to you. Joe said, "Try to find the middle ground," and I said, "There is no middle ground with me."

July 20 – The new boss barked at me again today because I went to the shop to get a tool without his permission. I cursed at him and told him I have a mind to knock his teeth out. He said he's the boss and get used to it and I realized I could not in good conscience hurt him because he's too old to fight effectively.

July 23 – I quit my new job in Pilot Station today and flew to Bethel and on to Fairbanks in the afternoon because I can't find anyone to fight and that's the only way I know how to settle anything.

July 25 – It is 11 p.m. and I am nipping on a bottle of whiskey in the Fairbanks Hotel. I am right back where I started two months ago before I took my first job on the *Pat* and it is not a good sign as I have now quit two jobs because of my temper.

A few minutes ago two native guys were shoving each other in the parking lot. A dirty white man with a beard and long hair showed up from behind a dumpster and began a tirade of repetitious nonsense. When I yelled out the window and told them to shut up, the three derelicts staggered off waving their arms and gibbering to each other in the cool midnight light of Fairbanks.

August 2, Tuesday – I went on a drinking bender last week and this lifestyle has finally caught up with me. On Thursday I am required to appear in Circuit Court as I was arrested for fighting and disturbing the peace last night. When I was in the Savoy Bar a cute native girl sat down next to me and said, "Buy me a beer." She did not introduce herself which is a common theme in this place and after I bought her a beer a guy at one of the tables threw a drink on his friend and some of it got on me. I took offense and threatened the guy. We went outside to settle the problem and one punch led to another until a police officer shot me in the face with pepper spray. Maybe I should have managed the situation without getting angry but when I am drinking in bars it doesn't take much to set me off.

August 5, Friday – Yesterday I appeared in Circuit Court with bad feelings toward the justice system. I think a man should have the right to a fair fight without interference from the law but I held my temper under control and received a sentence of thirty days in jail that was suspended by the judge provided I don't get into any more trouble.

I want to stop drinking and get back on the river, so today I drove to the river port town of Nenana and applied for another job with Yutana Barge Lines.

I explained my problem with Captain Murphy to the Operations Manager and he said barge work is always a challenge and Captain Murphy knows the lower river and gets the job done. He hired me back and told me to report to work in Nenana tomorrow night at 9 p.m.

August 6, 9 p.m. – I have just arrived in the Tanana River port town of Nenana as a deckhand for a tugboat named *MV Rampart*. It is a cool evening with a light breeze that carries an earthy, wet scent of the river.

The first mate is glad to see me as he is drunk, has two women and wants to go to the bar in town. He says I must stay on the tugboat to provide security and I will work the forward watch from 6 to 12 o'clock every twenty-four hours.

"Don't let any drunks on board," he says. "If you have any trouble, pull the chain in the pilot house."

He is referring to the possibility of problems with unauthorized persons or local native Indians who sometimes get drunk and come on board. He says pulling the chain will sound the boat horn and alert the rest of the crew that I need assistance.

I worked half the night installing a depth finder called a "fathometer" on the bow of the smaller of our two barges. The current swirls and murmurs past the hull in the dark and I look out over the river that will take me away from my troubles with alcohol.

August 7, Friday – I woke up at 5 a.m. when compressed air started the engines with a loud shriek. We tackled up with our two barges and headed downriver at 6 a.m. as predicted by First Mate Curtis. Today I lowered Pilot Casey in a small "kicker" boat, said "Good morning" to Captain McCarthy, coiled rope, pulled cable, tightened winches and moved equipment around on the barges with a forklift.

August 8 – We arrived at the Yukon River village of Galena in the night and unloaded diesel fuel at two sites. When I threw the heaving line to the native guy on shore, the weighted knot called a monkey fist went between his hands and hit him in the forehead which knocked him off a pile of driftwood into the river. At that point I observed that he was drunk and I had to jump over to shore and help him haul the barge's bow line up to the "deadman" which is a post in the ground used for securing the barge.

This operation in Galena was much easier however than our encounter with the village of Ruby when we worked for a solid hour just trying to line up the two barges so we could secure ourselves to shore and unload our deck freight. Most of the operation was accomplished with the port, starboard and center winches on the bow of the tugboat. They are hand-operated winches which is a good workout. The loading ramp is steep, muddy and a mess. One forklift blows so much smoke it looks like it's on fire. Driving either forklift on the barge when we are underway is weird. I feel like I am going over the side and tend to lose my sense of direction when I watch the river sliding past the hull. Our barges are so heavy with deck freight and fuel they look like they are more *in* the river than on it.

When I got off duty in Galena I went to a dance in the community center which is an old log building. There were about a dozen native elders dancing to records from the 1950s. I went to Hobo's Bar and had a few beers and stayed until 2 a.m. I felt like I was a sailor with a night on the town.

The next day three Christian missionaries asked us to put their boat on the barge and take it back to town for repairs. Their engine conked out. "God's Fish Wheel" is painted on the hull along with the boat's name which is *The Roland Lord*. These people want to bring Jesus to the natives that live along the Yukon River. Good luck to them. They are going to need it.

August 10, Monday – I think it is Monday but I am losing track of time while working and watching the river coasting by on its long journey to the sea. We left the Yukon today and are on the way to our home port of Nenana. The hills we passed this morning were covered by a thick blanket of fog and a few low clouds until the fog lifted to reveal a bright afternoon. We passed the *MV Tanana* in the narrows of the Tanana River. She is the largest tugboat in the fleet of Yutana Barge Lines. We waited outside the channel and she steamed past off our port side. A very large man wearing a dirty T-shirt waved at us from the bridge of the boat. I correctly assumed he was the engineer as all the engineers I have seen so far have been large persons who wear dirty T-shirts. Our captain blew the horn twice in reply and we continued across her wake upriver.

August 11, Tuesday – Today I pumped bilge water from the air void on the starboard bow of the portside barge. This section provides buoyancy and is a dark, wet place where I could feel the river hissing and slapping against the hull. I did not like it in there and did not want to stay any longer than necessary. We are making good time moving up the Tanana River and should be in Nenana by 2 a.m. Wednesday morning.

I am off duty now so I'll be in my bunk when we tie up in port. I share this room with Nate who is the other deckhand and assistant to the engineer. We will replace each other every six hours. I saw my first electric yard light a little while ago and it is evidence that we are getting close to the "civilized" world again after passing Indian fish camps, brush-covered bluffs and dark spruce woods for the last five days. I stood part of my watch on the bow of the starboard barge listening to the Sandhill cranes and Canadian geese taking off into the sky when they hear us coming around the bend. It feels good to be back on the river again.

August 13, Thursday – We arrived in port yesterday about 2 a.m. After cutting loose our two barges which is referred to as "untackling," we secured them to the freight dock and put into shore across the road from town. On this night everyone on the crew went to the bar except me because I had the forward watch. Moocher's Bar will be open until 5 a.m. and in the morning everyone will have a headache. I had a great time sitting up in the pilot house eating everything in sight, listening to Mozart on my tape player and watching the sun set over the highway bridge. This association with classical music calms me down and reminds me of my parents and the stability of my childhood home in North Carolina.

A big boss from Seattle stopped by earlier and almost fell in the river trying to get on the tugboat. He said he wanted to shake hands with the crew and I told him they were all off duty and in the bar so he decided to go back to his hotel in Fairbanks. He was friendly but he did not understand this work or have the slightest idea about where we were going or where we had been. He said 'Don't fall in the river or do anything dangerous' and I found this advice ridiculous.

August 14 – Casey has been far out in front of us with the pilot boat today marking the channels that change every day due to the heavy silt content in the Tanana River. We need at least four feet of water and after consulting a fathometer he must use a five-foot wooden stick to confirm the minimum depth. When he has found a channel he drops a red floating marker with a rope attached to an old starter motor or alternator which serves as an anchor. Several times during the day I could hear the hull scraping the bottom and, after almost running aground, Captain McCarthy pulled in to the north shore to wait for Casey to locate the channel.

I jumped off on shore and watched the Sandhill cranes fly by in formation. They are gathering together in large flocks to go south for the winter along with ducks and geese. The river is alive with these waterfowl on the move and it makes me restless too. What will I do with my life now? Thoughts of the past and future bother me and I don't want to think about it. I like the river and only the present matters out here.

It is now 10 p.m. and one of my favorite times on the river. Captain McCarthy decided to pull over and tie up for the night as it is getting dark, and the channel ahead just before we reach the Yukon is in bad shape with low water levels. He is tired and wants to attempt it in the morning at daybreak. He made his turn at a wide place in the river and headed back against the current, as this procedure provides the most control due to the resistance against the rudder. We then pulled in to the cut bank on the port side.

I dragged our bow line on shore, secured it with a shackle by wrapping it around two big birch trees and spent some time standing in the woods at dusk, watching the river and hoping to catch a glimpse of the full moon. I identify with the moon and think of it as a friend in the night and am always touched by its different phases and position in the sky. The clouds are lit around the edges with a silver glow and this has been a fine evening on the river.

It is 6:30 a.m. and we are attempting to navigate Squaw Crossing. It is the worst place for boats on the river as it is a confusing tangle of sandbars, snags and fast, shallow water. The crossing is a procedure by which the barges must transfer from one side of the river to the other to remain in the channel. There are no navigation markers on the river and the boat operator must use a twisted tree, large boulder or some other natural landmark to show him where to make his turn.

    Today we ran up on a sandbar and got stuck several hundred yards from the eastern shore. Casey was out in the pilot boat trying to locate the channel but he could not find enough water and it was almost three hours before we were able to wash out a channel with the tugboat's props and get moving downriver again. I spent most of my time sounding the river with a long, wooden pole and yelling out depths to First Mate Curtis who relayed this information to the captain. I felt like I was Mark Twain in the old days. We had to put on life vests which at first are a little uncomfortable but it was a cold day so at least they were warm.

    It is midnight. We are in the village of Tanana now and I am off duty which is why I can write. I still like this boat.

August 15 – It is Saturday night and we were still 45 minutes upriver from the village of Ruby when the moon rose above the hills at about midnight. The moon had a hard time moving above the dark clouds on the horizon. It suddenly appeared just as we passed close by the shore along the west bank and the sound of our engines started the dogs in someone's fish camp to howling. It was us or the moon that caused it. I also believe that a dog's hunting instinct is aroused by the moon. The Indians had a campfire burning and a glowing wall tent. It looked warm and friendly and I wished I could stop in and visit with them.

We kept moving and made it to Ruby but did not unload our freight because the ramp was steep and muddy due to a low water level in the river. I stomped through the mud and darkness for a while trying to move a boat that was in our way until the captain said "Let's go. We'll unload on the way back."

First Mate Curtis and I were sure glad to hear it as our barge lights did not work either. A man who said he was the mayor of Ruby came down to the river and was drunk. He ordered us to unload our freight anyway but we ignored him and were underway to Galena early Sunday morning with a big moon shining on the water and our engines at full throttle. I am writing in my bunk and it is nice and cozy in here. I am getting mixed up on the days. I thought it was Sunday but it is Saturday. The throb of our big Detroit diesel engines used to keep me awake and now they just lull me to sleep.

On Saturday night I went to Hobo's Bar in Galena after I was relieved of my watch at midnight. I walked around for a while after I got to shore and made it to Hobo's about 3 a.m. and had fun drinking Olympia beer and reading the walls. Someone wrote: "You only live once, but if you work it right, once is enough." While I read it I listened to the juke box and Jerry Lee Lewis singing "I Wish I Was Eighteen Again." I thought about it and decided I am glad to be thirty-four.

An elder native man came up to me and told me that beaver meat is very good and the tail is the best part. His nephew came up and asked me if I knew what Olympia meant. He said it meant "Oh, Lord, why make poor Indians alcoholics?" He also said he used to work on the barge and we had a good time talking about working on the river and the challenges involved. We talked about the different villages and he said everyone always looked forward to the arrival of the barge and that I would be welcome at the annual *potlatch* which is a native festival where everyone brings food and celebrates their traditions of friendship.

August 17 – It is Monday and the sun came out shining on the hills as we departed Ruby today. The Kokrine Hills have new snow. It is 1 a.m. and I am off duty. We are slowing down and will be taking on fresh water in a few minutes from a creek that enters the Yukon from the north. The full moon is shining on the water. The stars are out and it is a beautiful, clear, cold night on the Yukon River.

When I woke up this morning at 5:30 we were stopped on shore somewhere and tied up due to the dense fog. I stepped out on deck and could barely see the river. There was a hint of blue sky above us and the promise of a fine day when the mist lifted. In several hours it began to burn off and we backed off the shore and pushed upriver among the rising vapors.

As we came around the first bend I watched a bull and cow moose step out of the willows on the starboard side, touch noses and swim the river ahead of us. Captain McCarthy throttled back to slow down and the two moose lovers climbed out of the river on the other side and vanished together into the spruce timber.

Due to the low water level in the river I have to go further with the cable to secure the barges and I feel like I must have dragged it 500 yards back and forth through the mud yesterday. It must have taken ten cups of black coffee to bring me out of my sleepy haze this morning.

After lunch I went to the wheelhouse to visit with Casey. He said the river is always changing and you never know what will happen around the next bend or in the next village. He had just received word over the marine radio that the *Tanana* is stuck on a sandbar in very swift water at the mouth of the Tanana River. They are waiting for us to help them get off it. Casey says maybe we can wash out an escape channel with our engines or attach our big hawser line and pull them free of their predicament. They are just off Mission

Island and we will be there in about three hours.

August 18, Tuesday morning – It is now 1:30 a.m. The northern lights are out along with a big white moon and the stars. The sky is so bright it is difficult to see the light show and even at this hour there is still a faint orange glow in the northwest where the sun disappeared. The *Tanana* got off the sandbar by washing a channel with the props and we are moving down the Yukon River in the moonlight.

At noon today we met the *Independence* at Tanana village. She is the smallest tugboat in the fleet and has been waiting for us because she does not have enough power to buck the strong current at the mouth of the Tanana River right now. We left our two barges in Tanana and pushed them through Squaw Crossing, then left them to continue along the north bank alone. Without our barges to act as a deflector, our tugboat pushes so much water it pours over the bow. After retrieving our two barges in Tanana we headed back upriver and passed the little *Independence*. They followed us for two hours but were having a rough time of it so we tackled up to them. Now we are running together side by side through the night with their Captain Jensen and our pilot, Casey. The *Independence* is on its way back from Norton Sound and the ocean and they don't know the upper river well enough to travel alone in the dark.

Early this morning about 3 a.m. I woke up suddenly when we hit a sandbar. We were traveling fast and broke tackle with the *Independence* when one of the forward snatch blocks flew apart along with all the cleat and breast lines. Casey and Captain Jensen got the boats stopped quickly but I heard it was a mess for a while trying to get tackled back together and off the sandbar. I was not on duty and no one called me to help so I went back to sleep and surveyed the damage with a critical eye in the daylight.

After leaving Tanana village yesterday I noticed a long line of fish wheels tied up by the Indians along the riverbank. Fish wheels are large wooden and wire mesh scoops on a raft that use the power of the river to catch salmon. The salmon runs are just about finished for the season and chum, called dog salmon, are about all that is left. A few native fish camps still have salmon drying on the racks but nothing like a few weeks ago when every camp was full of fish.

Presently we are cruising past brown dirt bluffs that are over a hundred feet in height. I am standing on the bow of the portside barge where the engines make a soft purring sound. The wind has increased, the sky is clear, and long, thin clouds are lit around the edges with a subtle and beautiful shade of pink. The sun has set behind the Kokrine hills as they stand in sharp outline against the northwestern glow of this late summer afternoon. It will be a fine evening on the Yukon River.

There is more to the bluffs than I just mentioned. Casey told me that the bluffs are the bone yard of prehistoric beasts and many skeletons have been found and dug up in the vicinity. As we passed on I was sure I could smell their rotting flesh. The odor was very strong and I thought it smelled like the zoo.

I went up to the pilot house to visit with Captain McCarthy after dinner. He told me some good stories about working the lower river at the mouth of the Yukon and Norton Sound: When he was a pilot, a drunk Eskimo captain with a knife chased him down the beach in Nome. When the Eskimo stumbled he knocked him out with a piece of driftwood.

Another time he said he was heading down the coast at night with one barge in tow when the boat started taking on water. He ended up off Cape Darby with the engine room flooded. After the necessary repairs and pumping out the seawater, he made it to the Yukon with the tide and, after numerous attempts to navigate the maze of channels at the mouth, he made it back to our home port of Nenana in three weeks.

On another occasion he had to hide behind Besboro Island in a storm and he said he was glad to be out of the ocean and that he preferred the upper river.

August 21 – The night is dark and full of noise where I am lying down on the stern of our boat above the engine room. The metal deck beneath me is vibrating and warm from the heat of the engines and, despite the pounding of sound in my brain, this powerful, steel vibrating bed feels good to my aching back. Black smoke pours out of our exhaust stacks, then flattens out low and retreats fast over the roiling river behind us in the night. A yellow half-moon is creeping up over the hills. This evening is full of mystery. The moon looks like the eye of the universe and there is a feeling of being watched.

August 23, Sunday – I am standing on the edge of a high cut bank above the muddy Yukon River in Galena. The fuel transfer pump is roaring at 1500 rpm. It is pulling 110,000 gallons of fuel oil out of the barge bulk tanks and dumping it into the U.S. Air Force shore tanks. The grass and leaves are bent to the north and a south wind is forming a heavy chop out in the middle of the river. Once in a while a covered flat-bottom riverboat with a native family aboard scurries past on their way to and from a fish camp or to visit relatives in another village. These people use boats like the residents of cities use cars and the river is their highway.

Monitoring the fuel pump is my job here and I have to shut it down fast if a leak in the fuel line develops. It is very cold on top of this bluff above the Yukon River and we have already been here for sixteen hours.

When I got off duty in Galena last night I went to Hobo's Bar and had a conversation with a white guy who changed his life considerably. He is a California surfer who just married a native woman he met in Fairbanks. He says he will live in the village and extended an invitation to come over to his house sometime for a meal of moose-head soup which is widely considered the best soup on the Yukon. I wished him well and he went home. His departure evidently inspired a very large woman from Anchorage to sit down next to me. Then a cute, young girl came over and asked me if the large woman was my wife. I was amused by this question and made it clear that I was not married because I hoped to see the pretty girl again. Of course I would depart down the river in the morning with no chance to get acquainted with her which is the life of a deckhand.

The sunrise this morning came on fast and started the day with a blinding brightness reflecting off the water. We are now underway with no end in sight to the huge mass of clouds moving in on us from the south.

August 24 – Today we passed a young black bear who was ambling down the long, straight gravel shoreline. He heard us approach and ran up into the timber where he stopped and watched us as we proceeded.

Late this afternoon we passed several abandoned cabins that are dark and silent in the birch timber. They seem austere to me and I wonder who loved, laughed, cried and played in this meadow and under the trees way out here along the Yukon River. Now and then we pass a group of cabins with boats tied to driftwood along the shore. Once in a while someone will step outside and wave to us and I return the greeting. There are very few people who live out here and to acknowledge someone's presence is almost a duty. It is something that is enjoyed in its simplicity for a friendly wave is all there can be under the circumstances.

On this night I went up to the pilot house to visit with Casey. No lights are allowed in the pilot house while underway as they would compromise the boat operator's night vision. I looked down at the shadows of deck freight on the barges and out to the reflection of starlight shimmering on the wide stretch of dark river ahead of us. Casey asked if I had any plans for the winter. I told him 'I don't know what I will do'.

Time passes like the endless shoreline and unless I think about the date I lose track of it and the passage of time is not important. The changing season and natural rhythm of daylight, darkness and river conditions dominate our mission of delivering freight to the people on the river.

The tugboat's rudder shaft has developed a leak that is becoming worse. Pilot Casey said the leak has been there all season and no one has been inspired to fix it as we have not yet begun to sink as a result of it. When we reach that stage someone may be interested. He says that unnamed "someone" is a large person who wears a dirty T-shirt and is responsible for maintaining the mechanical integrity of the tugboat. This person, of course, is engineer Kevin whose favorite activity is dispensing orders to the deckhands. He yells at Assistant Engineer/Deckhand Nate sometimes but does not yell at me because I told him I am a boxer, even though my boxing experience was acquired from altercations with drunks in the bars of Fairbanks. First Mate Curtis is my boss and I get my instructions from him. Mostly I think we have a good crew, even though there is some discord among those that live and work together so closely. There is nowhere to go when a fellow gets mad unless he wants to jump in the river. I go out on the barge when I need to be alone to collect my thoughts and calm myself.

Yesterday, while pushing two barges up the Tanana River, here comes a log raft with a wall tent and two scroungy, long-haired fellows frantically trying to avoid the approaching barge monster with the use of two long poles. Unfortunately, they steered into the main channel between the riverbank and the barge which can be a mistake if you want to stay alive. They had ten yards or less and they squeezed past, losing logs and struggling to stay afloat. The *Rampart* generates a monstrous wake and, as the raft plunged astern of us, it dipped in and out of sight before being lost from view around the bend. Just before they disappeared, all I could see was a heaving mass of logs and hair and I wondered if they survived. If they did survive and live to be old men, I believe this adventure will be in their conversation for many years.

Later in the day we saw three black bears on a long steep ridge above the Tanana River. They watched us without fear as they were far above, looking down with small eyes and listening with little round ears while no doubt using their noses to determine what we were. I like to see them but I don't trust what they would do if I had a close encounter. Casey said young black bears who have never seen a human can be especially dangerous as they have not yet learned to be afraid of people and they regard humans as prey. He says these bears are the street punks of the wild and the ultimate opportunists. Casey says he carries a large caliber rifle or 12-gauge shotgun for protection when he is in the bush.

August 28 – We arrived in our home port of Nenana yesterday afternoon and when I was relieved of my duties at midnight I drove to Fairbanks. I went to the Savoy Bar on Second Avenue because I had been there before and the native girls were friendly to me. They were mostly straightforward in their approach and I did not have to guess at their intentions. I drank beer and listened to "Bad Moon Rising," "Rolling on the River," and "Islands in the Stream" on the jukebox. I know the natives could relate to these songs because most of them came into town from villages along the river and I believe they feel like "islands in the stream" themselves. When these people come to the city they enter a much different world from their quiet little villages along the river.

After about twenty minutes in the bar a native girl sat down next to me and asked if I wanted some company. I said I would like that and I bought her a beer. One beer led to another until I ended up across the street down by the Chena River in the willow brush, drinking whisky from a bottle with her until two friends from her village of Ft. Yukon arrived. At that point it was time to drive back to Nenana as I had to return to work on the boat at 6 a.m.

During my drive back to Nenana I made a conscious attempt to be careful after having been reminded by Pilot Casey that I had replaced the deckhand who was killed in a wreck while driving back from Fairbanks drunk at night.

August 29 – In the morning at 7 a.m. the big tug *Tanana* pulled around our boat to make up her tow of two loaded barges. The crew was standing on the bow and with smoke pouring out of her exhaust stacks and laundry flapping in the wind she looked like a riverboat from the old days. I think these are the good old days up here right now. When our barges were loaded at noon with bulk fuel and deck freight we edged out into the current, swung around with our stern almost touching the docks and headed downriver. I could see big old Mt. McKinley standing tall and bold off to the south against the midmorning icy blue sky as we passed under the silver highway bridge. We waved goodbye to people watching us from the road as we slid sideways around the first bend and out of sight. After one night in town it sure felt good to sober up and get back on the river again.

Driving the tugboat is an art. It has to be turned at precise moments, wrenched at times and coaxed along at others. Moving downriver with loaded barges takes a great deal of looking ahead and planning the next move as the current is the center of all power here. The boat operator must use it to steer around sharp bends. The navigation of a sharp bend is obtained by nosing the barge almost against the shore on the inside of the channel, reversing engines, backing out into the current and crabbing sideways with the river around the bend until the tow is straightened out, and then pouring on the power to push around the inside of the curve and forward on down river. The boat operator must also read the water to stay in the channel by looking out for riffles and slack water that indicate the presence of hidden sandbars, and he must contend with the effects of wind and watch out for snags and drift logs. Today we passed the confluence of the Kantishna River where its color and personality swirls out into the Tanana River increasing its volume and velocity.

Our air compressor is too small and faulty. It is also one of the most critical machines on the boat as all rudder control is powered by air pressure. The engines are started with compressed air and I will never forget that sound, which is a loud moan, almost like the shriek of a wild animal, and then pounding of the pistons begin. We lost air pressure and crashed into the cut bank three times today. On one occasion a snag came through the open window of the galley and almost took out the cook and the engineer who were sitting at the counter. This trip began with problems. The river level is dropping every day and I feel like our troubles will get worse.

August 30 – This morning there was almost a foot of water in the engine room and we had to use the fresh water pump for bilge because the bilge pump quit. Last night we pulled into shore below Swan's Neck Crossing and after I secured the bow line to a big spruce tree I spent part of a cold evening standing in the woods under the stars and brilliant northern lights, just outside the glare of the tugboat's spotlight and thought about my girlfriend. I knew she would be with someone else and the thought made me crazy. I kicked the ground and threw sticks into the woods but when I looked up at the lights pulsing in the sky from north to south in a cold white band of mystery, I realized the problem with my girlfriend in Oregon does not matter anymore and I have to let it go. The small points of the Big Dipper were shining through the mystical haze when I returned to the boat with a feeling of considerable relief and went to bed in my little room behind the forward bulkhead.

Today we hit the treacherous confusion of sandbars and shallow channels at Squaw Crossing with a vengeance. She's hurt us too many times. Captain McCarthy was determined to ram our barges through a narrow channel but the Tanana River accepted the challenge and fought back. Pilot Casey marked two of the worst places with buoys but we couldn't make it and got stuck on the first bad turn. I was off duty but brought into service quickly to aid my comrades in their efforts to free ourselves. The current was swift and I was reporting only two or three feet of water along most of the port side channel which was the direction in which we would have to escape and whip this tortuous maze. Captain McCarthy decided to use the engines to wash out a channel with the props and during this procedure sand began to heave up out of the water alongside the tugboat in large mounds. The engines roared and the mounds grew larger and looked almost like monsters from the deep that were thrusting their heads up from the bottom to devour us for daring to disturb their domain.

When this approach was not successful we were instructed by the captain to jackknife the barges to the port side which allows the barges to swing out in the current with the expectation that they will jerk the tugboat off the sandbar.

Upon the captain's signal to First Mate Curtis I slacked the starboard winch line, Engineer Kevin tightened the port line, Assistant Engineer Nate operated the center and Curtis acted as a communications link to the captain. We were set to break free with the barges jackknifed in the portside channel and all hands on the winches ready to swing her back when the river took on a mean personality.

The barges slid off the sandbar really fast and we were not able to get the barges tackled back to the tug before the river took control. The center winch disintegrated with Nate on the brake and he was slammed against the bulkhead. He ran into Kevin trying to escape flying metal as the entire port side winch line ripped free of the drum and out into the water. I slacked the starboard winch and jumped clear to let her go which is the only thing I could do. The boat heeled over violently sideways and seemed ready to capsize while slamming into our two barges that were beginning to escape from us.

Curtis and I leapt across to the escaping barges with the only lines we could grab to hold them. We secured the lines to any cleat or post that was handy to us and they stretched, groaned and held as the tug reeled drunkenly in the swift current sideways to our tow of barges and out of control. The river banged us around without mercy and toyed with us like we were insignificant pieces of drift.

Casey sped out of the way in the pilot boat and we began hauling all the cable we could find up out of the forepeak. We used everything we had to replace our lines and secure our barges before we lost everything. It was tough but we prevailed in our efforts and I knew our situation was serious but I did not panic. I was also glad no one was killed and I wondered what other surprises were waiting for us on this wild and unpredictable river system.

August 30 – We unloaded supplies for a homestead family at Mouse Point today. They have a cabin about a half mile down the trail in the spruce timber. Casey knows them and said they get around by dog team in the winter and make their living on the land by hunting, trapping, and growing much of their own food.

While hauling 50-pound bags of dog food, potatoes, rice and other staple foods on my back off the barge on wooden planks to shore, then toiling across the riverbank mud to a dry spot at the trailhead, I felt again like I could be unloading a riverboat in the old days.

The engine room is flooding and becoming a large problem. Engineer Kevin is in a perpetual rage about it as he must crawl down into the small space above the rudder and change the seals while lying in water, mud and grease. It will do him good because he needs to do some work as he is quite large and he has been sitting on his butt most of the time anyway, eating and leering at our cook, Hazel, in the galley.

In the afternoon we set off some freight at Big Bend gravel bar for the gold miners that are working a claim up in the hills and later we arrived in the village of Ruby at sunset. As usual I had to move a herd of small boats out of the way so we could pull into the loading ramp. Several drunk natives saw me untying their boats and came lurching down the hill yelling at me to leave their boats alone. I said "Help me move your boats so we can unload our freight. I don't have time to talk about it."

One man got into a boat without a motor and the other man shoved him out into the river. The guy in the boat started yelling that he was going to be lost in the river without a paddle so I jumped into the water and pulled him back to shore while the other guy staggered and fell in the mud and was of no assistance whatever. It was a great chore hauling cable through the mud to secure the barge's bow and stern lines. Of course I also had to yell at the howling crowd of sled dogs that were tied up between our barge and the post which I must use to secure the bow line. I had to drag the cable through the dog lot and I cursed at them because they were snarling at me.

    We worked most of the night in Ruby unloading fuel and mining equipment and dragging fuel hose 100 yards up through the mud to the fuel truck that is so decrepit that it cannot be started and moved down to the barge which would make the job much easier for us. A native guy who manages the village power plant watched the shore tanks for me while I patrolled the hose line for leaks and operated the pump. He was very helpful and he told me about a friend who was killed last winter in a plane crash at Galena. The aircraft was overloaded and it stalled on takeoff and crashed on the river ice.

The manager of the power plant also told me that another friend got drunk after a trip to Last Chance Liquor Store on the Yukon and fell out of his boat near the bluff at the village of Koyukuk. That is a bad place as the river channel narrows and sweeps around the bluff with a strong current. Searchers found his body way downriver near the village of Kaltag. The manager said he himself had stopped drinking two years ago and the change made life much better for him.

This night is cold and clear and when I shut down the fueling operation I stood on top of the Shag Mine fuel truck alone and looked up into the black sky full of stars and I wondered if there could be a deckhand in another world somewhere out there billions of miles away looking back at me.

September 1 – We nosed into shore at the village of Koyukuk tonight. It is a small village just off the main channel of the Yukon and upon three blasts of the boat horn to announce our arrival, all the dogs began to howl and many people came down to the loading ramp including some young girls.

I jumped off the barge as it touched the shore and I was watched by the crowd as I climbed the cut bank and dragged the bow line through the woods. The girls were whistling and calling out to me from the willow brush and behind the trees. The tugboat's spotlight was shining on me and I could not see their faces but when I answered they giggled and laughed and it was all a lot of fun.

We brought many items to these folks that had been ordered for at least six months which included snow machines, all-terrain vehicles, boat motors and household appliances. The people were excited and I felt like Santa Claus of the Yukon. First Mate Curtis dropped an electric cook stove off the forklift by accident into the river where it will remain for eternity. We will bring the customer another stove on a return trip.

September 2 – Today we are on our way with one barge up the Koyukuk River to the village of Huslia. The Koyukuk is a clear amber color. Just downstream from the mouth of the Kateel River we were passing a native woodcutting camp when two boys jumped in their boat and took off. They cut across our bow twice very close and this behavior made Captain McCarthy angry due to the great danger involved. He gave them a long blast on the boat horn and the boys sped away. Captain McCarthy told me later that a few years before when a small boat got too close, the barge ran over it and the occupants' bodies were never found. Small boats should give the barge plenty of room.

We pushed our barge up the Koyukuk all day and I watched a black bear nosing along the gravel shoreline at sunset. I secured the barge to a spruce tree on shore for the night as our captain does not know the river well enough to travel in the dark. I climbed the cut bank at midnight and walked off into the woods away from the spotlight and sound of the generator where I observed the Aurora Borealis playing across the sky in long, pale white and yellow streaks of light.

A pack of wolves howled to each other somewhere along the riverbank and that mournful sound reminded me of the constant struggle of life and death that was going on all around me. Somehow I feel connected to it and when I am alone out here like this it is my favorite time on the river.

Earlier this evening while moving down the Yukon Captain McCarthy told me about the infamous Captain Bruno. This fellow seems to have run into every obstacle on both the Yukon and Tanana Rivers including the Nenana River Bridge. Bruno's Rock is well known to all barge men as Bruno rammed it at night and did considerable damage. He is described as a big hulking Polish man with huge sloping shoulders, no neck and a heavy square head. Pilot Casey told me later that Bruno is a friendly fellow who is not bothered by his reputation and is rather proud of it. I have not met him yet but hope to do so and engage him in conversation some day.

September 3 – We made it to the village of Huslia late this afternoon and had to push up a ramp with a dozer before we could unload our deck freight.

At dusk I worked the tanks on shore while we pumped fuel oil for the village. A native girl that does the books for the fuel farm came out of the office to talk to me. I climbed up and down the big rickety fuel tanks checking the fuel levels so I could close and open the appropriate valves to fill the next tank before the first one overflowed.

I watched the sun go down from the top of one tall tank and felt the instant bitter cold settling in on the north land night. The stars appeared slowly and the sunset's soft orange glow in the western sky remained a long time as a few last rounds were fired off by the natives at the ducks coming in to land on the pond just a few hundred yards through the trees.

After filling the village tanks we moved the barge downriver to the center of town by the big bluff to pump the school's heating oil.

Huslia is surely the most austere village I have yet seen. The landscape is stark and forbidding. It is also strangely appealing because of its proximity to this wide wilderness of stunted, twisted, cold country trees in a vast solitude.

Securing the barge down by the enormous bluff at the center of the village proved to be a very difficult operation. After hauling the bowline straight up the 30-foot cut bank I was compelled to drag the cable by inches on my knees through 50 yards of thick brush and a pack of howling sled dogs in the dark to the one tall, lone white spruce tree. All the while I cursed my plight out loud with every breath and unknown to me the native girl I had met at the village tanks was standing shyly off to one side of the tangle of brush listening to me. When I secured the cable to the tree and emerged from the brush I apologized for my bad language and she offered to meet for a visit on the bluff after I was relieved of my watch by Assistant Engineer Nate at midnight. He was late and when I finally made it to our rendezvous she had disappeared. I gazed up into the cold, clear, black sky full of stars and felt sad as I sure needed the company of a girl that night because I was lonesome and far away from home.

September 4 – Upon leaving Huslia in the morning we loaded a small front end bucket loader and a dump truck with two men from the village. We hauled this extra freight up the Koyukuk about ten miles to a gravel pit. The men will return during the winter to load gravel and drive back to the village on the ice. On our way upriver we passed a homesteader and his family who were standing on shore in their fish camp waving their arms wildly. Captain McCarthy blew the boat horn and we were greeted with a renewed surge of waving and excitement. Casey told me the homesteader left his wife in California for a new life in

Alaska. He married a native woman and now has an extended family of kids and relatives in the village. These people looked to me like pioneers of long ago.

September 5 – At midnight on our way back up the Yukon River we slid into the low gravel shore at Herman the German's fish camp between the abandoned community of Kokrines and the village of Tanana. The German's real name is Helmut Schmidt but the barge crew refers to him as Herman the German. He was asleep when we arrived and was not happy to be aroused on such a cold, dark night by the blast of our boat horn and the wicked stab of our spotlight focused full force on the door to his cabin. He lurched out into the blinding glare as First Mate Curtis and I watched him from the deck of the barge. His dog ran off yelping into the night as Herman kicked him and lumbered down the hill in our direction, cursing the dog, the cold and the barge for having awakened him at such a rude hour. We calmed him and moved his boat so we could nose in closer to shore and load his ancient refrigerated truck full of frozen salmon onto the barge. The truck would not start so we towed it aboard with the forklift which was also in constant danger of malfunction. The generator for the refrigeration unit on the truck sparked, coughed and sputtered all the way back to town but never gave up. We also loaded many bales of dried salmon and as numerous pieces fell out on the deck of the barge I ate a lot of dry fish on that trip.

September 8 – We arrived in our home port of Nenana yesterday and when I was relieved of my duties at midnight I drove to Fairbanks. I went to the Cottage Bar on Second Avenue in hopes of finding a girl with whom I could have some intimate contact. The bar was full of natives and white guys playing pool and chasing girls. After a couple of beers at the bar while I was observing the boisterous scene, a guy next to me hit his native girlfriend and started cursing at her. I told him to stop it and he cursed at me and extended an invitation to go outside and fight which I accepted. When we got out on the sidewalk he took a swing at me. I avoided the punch and knocked him down with a straight right hand to the jaw. Someone hit me on my blind side but as the punch lacked sufficient power to knock me down I swung around and caught him in the face with a back fist and as the two guys were crawling on the sidewalk and were no longer a threat I walked away. There was some yelling but I was not pursued.

The girl followed me and when she spoke to me I could see that she was frightened so I rented a room for the night at the Polaris Hotel for her protection. I did not attempt physical relations with her and when I said I had to get back to work in Nenana she asked me why I was so nice to her. I told her she deserved respect like everyone else and I drove back to Nenana with a black eye and a bad attitude. The crew noticed my bruise and everyone had a comment. Mostly they said I should stay in Nenana when we are in port so they could watch my back since I had a tendency to get into trouble in Fairbanks.

September 10 – We got stuck on a sandbar in the rain today about three hours upriver from Squaw Crossing. The river level has dropped considerably and is in rough shape for barge navigation. Captain McCarthy washed out an escape channel with the props and we continued downriver.

September 11 – Long streaks of thin red clouds preceded the rise of the sun this morning in Ruby and last night after Captain McCarthy went to bed Pilot Casey invited some native guys on the boat to watch a movie called "The Blues Brothers" on our video machine. The movie is about two troublesome brothers who plan to put their blues band back together to raise money for a Catholic orphanage. The movie is a comedy and the visiting natives really enjoyed watching the white guys acting like idiots. All of us laughed until we could barely breathe and it was a lot of fun.

September 12 – We arrived in Galena in the afternoon to transfer fuel for the U.S. Air Force and unload our deck freight. As soon as the crane operator on shore set a pallet of Olympia beer on the ground, several of his friends pulled a six-pack and began drinking it. The crane operator was also drinking beer. When I began pumping fuel from our barge to the shore tanks, the beer drinkers began arguing with each other and I told them to get away from me. They backed off but stayed in the area and I had to keep my eye on them as I did not want them to touch any valves. Turning the wrong valve or disconnecting something could result in a disastrous fuel spill. When they wandered off I went back on the barge to check the fuel line and spoke to Hazel. The noise of the fuel pump seemed especially irritating and I decided to see if I could extract an interesting conversation from her so I hollered, "There never seems to be any shortage of racket around here" and when she yelled back, "I can't hear you," I could see conversation was hopeless and gave up.

Hazel is not a good cook and she could cremate a hamburger and burn the toast but at least she is friendly and that goes a long way with me.

September 13 – We made it to the village of Nulato last night under a full moon rising at dusk. A crowd of natives greeted our arrival with much shouting and waving. They are excited to see us and we will transfer their village fuel in the morning.

This day began cold and bright while we unloaded our deck freight. The brilliant sunrise was welcome and I noticed the old folks and kids were the first ones to venture out this fine Sunday morning on the Yukon River.

It is midnight and the *Tanana* just passed us above Galena. She slid by lighted up like an ocean liner. We were planning to drop our pants and moon everyone on board but Captain McCarthy heard us conspiring to do so and warned the other boat. In that event we did not do it due to the loss of the horror of surprise.

September 14 – The sight of the Ruby village greeted us through the lifting fog at 11 a.m. this morning. There is the promise of a beautiful day. The fog is slowly retreating downriver and is silent and mysterious. I went ashore at noon when I was off duty and roamed around high above the river in Ruby.

This is the first time I've had the chance to go into town and get a good look at the village. The town was settled by gold miners who built their cabins on a hillside next to a high cliff. I walked past the Ruby Roadhouse which is a quiet little inn that seems unusual way out here on the Yukon. It looks very secure and friendly in its small glade of birch trees and reminds me of my boyhood home in North Carolina.

While I was in town the crew secured the barge with the bow pointed straight into shore for the purpose of loading a large dozer from the Shag Gold Mine. This operation requires a coordinated use of cable reels, snatch blocks and winches in a task of effort and complication which I do not like. We will transfer diesel fuel to the gold mine's tank truck and depart for Galena. I noticed the Ruby cemetery is on a hill above the village which is the custom of the native people in Alaska. The village of Nulato has a peaceful cemetery on a hill very much like the setting here.

Late in this night we took on fresh water from a little stream that rushes into the Yukon sparkling in the moonlight. I'll take a shower tonight as there is plenty of water and I will go to bed in the splendor of cleanliness as opposed to grimy resignation.

September 16 – Tonight we performed the battle of the searchlights as I operated one light while Captain McCarthy directed the other one and we crept our way upriver in the darkness. The specters of dark spruce and leafless birch appear along the shore and, as there are no navigational markers on this wild shoreline, the boat operator must find the familiar twisted birch, a tall spruce or unusual clump of root wads to mark our progress and show him where to cross the river in the blackness and remain in the channel. Claude Dementief is ahead of us with his little tugboat *Ramona* pushing one small barge. I can see his mast light on straight stretches of the river and we are gaining on him as we have more power but he will go almost anywhere in shallow rivers and haul almost anything to anyone.

September 18 – Yesterday we acquired a new air compressor and maintenance was performed on the engines.

It is now early morning and we just left Nenana and are swinging around with the current to straighten out our tow of two barges after backing under the bridge. Our barges are loaded with heavy mining equipment and I watch the cars moving down the highway in the rain. They pass out of sight as we slide around the first bend in the river that will take us away.

Tonight the half-moon looks lonely up there in a very dark and cloudy sky. The searchlight shines on our bow line that is secured to a large spruce tree up on shore. Two big yellow Caterpillar dozers are squatting on deck like cold steel monsters that are resting the immense power they possess. We are tied up for the night as it is again too dark and dangerous to navigate the constantly changing channels of the Tanana River. I am sitting in the pilot house alone listening to Kim Carnes singing "Betty Davis Eyes" on a tape player while looking out over the loaded deck and the river swirling past the barge in the dark.

Earlier I went out in the pilot boat with Casey to find a place for the barge to cross over to the north bank where we can tie up in the channel. He dropped me off on shore at dusk and I watched the barge from a new perspective as it was moving toward me in the dark like a gigantic prehistoric beast.

September 19 – It is a cold, wet morning where I stand my watch on the bow of the *Riverways* barge while looking out for snags and drift logs that we must avoid. The channel is definite here but the river level is low. We are moving along and Casey continues to speed past us in the pilot boat. He is laughing and drinking a beer. There is a skiff of new snow way up high in the hills and the crew is in good spirits as the navigation season is winding down. I feel some excitement and apprehension at leaving my simple life on the river for the winter since it has provided a respite from the disturbance of having to make decisions about what I will do with myself now.

A couple of moose hunters in an expensive aluminum boat just buzzed by the barge at dusk. They waved at us and Casey said the moose are probably feeding just behind the willows watching them and are unconcerned because the hunters are moving too fast and are conducting themselves like they are unaware of the habits of the moose who will hear them coming and disappear.

September 20 – When I woke up this morning we were tied up in Tanana while pumping fuel into the village shore tanks. Upon our departure we took off downriver to Lancaster Point and endeavored to unload the big dozer on a flat place at that location for a group of gold miners. This plan was a mistake as the dozer sank into the mud halfway off the unloading timbers. Captain McCarthy backed the dozer up on the barge again and we went about a mile downstream to more stable ground on a gravel bar. After using the smaller D-8 dozer to construct a ramp we unloaded the big D-9 dozer, the mammoth rocker mining machine and the juggernaut, which is a large tracked vehicle. I liked to sit in the cab of the juggernaut on my time-off duty and watch the changing scenery as we passed down the river.

During our unloading process the gold miners landed on the beach in two single-engine aircraft a quarter-mile away. In about a half hour we heard them crashing through the brush. There are four of them and they are real characters. The fat man is the boss and he says they need to get their ass in gear because they have a long way to go. One of them is an old man with a limp and a .44 magnum handgun in a shoulder holster. The most talkative miner is a little guy with a harelip and the fourth is the youngest who is wearing coveralls that are about three sizes too big for him. These guys are planning to walk their equipment eighteen miles back into the bush to a secret location they intend to mine next summer.

September 21 – It was a windy rough trip to Ruby but after a trouble-free docking operation we unloaded our deck freight with no problems. We also picked up a canoe for a guy from Minnesota who floated the river from the village of Eagle which is a trip of over 500 miles. We will take his canoe back to Nenana as he says it is late in the year and he is ready to go home. When I asked him about his journey he said that his canoe kept him intimately associated with the river and, when he was navigating the Yukon Flats below Ft. Yukon and he did not know which channel to take, he learned to trust the river, put down his paddle and let the river carry him through the maze of sandbars and shallow channels. He plans to return to Alaska next year and continue his float trip. We left him in Ruby where he will take the mail plane back to Fairbanks.

Upon leaving Ruby we left for Galena in the gray wet dawn and arrived there on a cold clear night with bad weather moving in from the west. The weather is not here yet but heavy cloud cover can be seen creeping towards us in the distance. We pumped our last fuel load of the season in Galena at 10,000 gallons.

September 22 – We cut loose from Galena this morning in a snowstorm and shot downriver to the village of Nulato. It snowed on the way and covered the riverbank with a white blanket. Winter is coming! Snow is here!

It is 10:30 p.m. in Nulato and we have had a rough day. We have unloaded a mountain of building materials for the *new* village of Nulato. At present the natives live in old log cabins down by the river which is very practical because the river is their source of food and transportation. A federal government subdivision is to be built on a hill above the river. Now the natives will need transportation to get *to* the river. Half the materials delivered earlier this summer are uncovered and have been deteriorating out in the woods.

Both of our forklifts are busted. Nate backed into a tree with one forklift and punctured the radiator. Kevin drove off into a ditch with the other forklift while frustrated with the situation and broke a steering line.

A wet snow has been falling all afternoon. It is dark and we are hauling the remainder of our large pile of deck freight and lumber off the barge on our backs.

Curtis and I are loading empty pallets onto our ancient flatbed truck which has unreliable brakes on a hill in the dark. We are all drinking beer and everyone has about had enough of this day.

September 23 – We got away from Nulato at 6:15 a.m. and on our way back upriver we tackled up to *MV Kantishna* about 10 a.m. Everyone is eager for new faces and new conversation so both crews exchanged information and visited one another until late in the night.

It is 11 p.m. now and we are underway to Ruby with a misty rain falling on the river in the dark.

September 24 – In the middle of the night we rammed the flat gravel shoreline ten to fifteen miles below Ruby right in front of someone's camp. It was soon evident by their accent they are a German couple on a float trip down the Yukon River and had the misfortune to be asleep when we crashed. It surely scared those poor folks. The man crawled out of their small dome tent to face the glare of searchlights, screaming diesel engines and a huge metal monster. Both captains throttled down just in time to avoid running them over. Because the bow of the barge on the *Kantishna's* port side was only twenty feet from the tent, I ran out to the bow of the big barge and was close enough to damn near shake hands with the stammering German. The glow of his smoldering campfire distracted Captain Jensen of the *Kantishna* and it was so dark he just got too close to shore. I hardly knew what to say to the camper so I said, "Sorry, we'll be leaving shortly." I felt like an idiot.

We made it to Ruby early in the morning.

September 24 – Sometime late in the night I secured our tow of barges to a big spruce tree after leaving the *Kantishna* in Ruby and heading upstream. I climbed a 15-foot cut bank with its overhang of roots and soil that is constantly breaking off into the river due to the action of the current. This precarious latticework of hanging snags and roots did give me something to grab and made it possible for me to climb it.

I was informed later in the afternoon by Kevin that Nate damn near busted his ass in the morning trying to climb the precipice to release the cable. Nate is not very coordinated and has fallen off the barge four times in five months. On one occasion he was pulled out in the nick of time after he fell in the river between the tugboat and the dock in Nenana during the arrival process. If he does not put his mind in gear before he moves his feet I hope he won't have an accident from which he can't recover.

September 25 – We picked up the *Riverways* barge at an island and left a pile of supplies at the mouth of the Nowitna River for a dog mushers' winter camp. I spoke to Captain Jensen of the *Kantishna* about navigation on the lower river and the ocean while I operated the searchlight for him at the Boneyard Crossing. He had heard about my problems in Fairbanks and warned me about the dangers of drinking in the bars on Second Avenue. I thanked him for his advice but I did not tell him that it is the unpredictable nature of such activity that makes it more appealing to me.

September 26 – We arrived in Tanana Village this morning and received word that the *Tanana* is grounded on a sandbar at Squaw Crossing. We have tackled up with the *Kantishna* again as she is underpowered for low, fast water and she lost an engine last night. Our port side engine is very sick. The mechanics are working them over in Tanana and they may be here all day and all night. The *Tanana* is still stuck and blocking the channel. Captain Jensen and our pilot, Casey, have approached them in the pilot boat to help them find a way back into the main channel and with a great deal of effort we were able to attach a line and pull the *Tanana* off the sandbar with the *Rampart* and get back to the village just before dark.

After leaving Tanana village at first light we made it through the maze of sandbars at Squaw Crossing, secured our two barges at the confluence with the Yukon and went back upriver to pick up the *Kantishna's* ocean barge.

It is now 7 p.m. and we are grounded on a sandbar above 8-mile slough with the little *Kantishna* waiting behind us. I returned to duty at 6 p.m. and our tugboat is slamming violently into the *Kantishna's* big ocean barge as we are desperate to get us both free of our confinement. When I walked into the galley, Casey's tape player came crashing down out of the pilot house and I was thrown against the bulkhead as the *Rampart* lurched around and barely missed ramming the *Kantishna* in the swift current. The galley is a mess of broken cookware and Hazel is frightened but we slid off the sandbar and I secured the barge to several large clumps of willows for the night. We used this opportunity to rest and make plans for our next attack on this river that is shallow, swift, ready for winter and tired of us.

September 27 – This day began with a gray dawn and the red glare of a rising sun to forecast the day ahead. There is the hint of a fine day but no promise of it. The pilots of both tugboats are up ahead sounding Squaw Crossing. Pilot Casey is working with Pilot Walker who is a native riverboat man that understands the subtleties of the river in special ways and can sense its dispositions.

The engines have been started and we are ready for another attack on that chaos of sandbars, snags and shallow, fast water one more time.

It is now almost noon and our tugboat with perseverance took us through Squaw Crossing with all four barges, one barge at a time. The *Tanana* and *Kantishna* are behind us and we are all pushing upriver together.

September 28 – I sounded the river this morning with my wooden pole after operating the searchlight at Patterson's Crossing last night. I will stay in Alaska and get another job at the end of the navigation season because this north country speaks to me and I have decided this is where I belong.

It is cold and clear tonight with a steady north wind. The spruce trees and willows on shore are all bent to the south. We keep moving and feel an urgency to get back to Nenana before the river begins to fill with ice.

September 29 – It is a windy morning at 25 degrees with silt blowing in long waves across the dry sandbars ahead. I worked the pilot boat with Captain Jensen through Campbell's Crossing, the treacherous turns of Goose Neck and all the way past the village of Old Minto. It is the end of the navigation season and we are on our way home.

Made in the USA
Las Vegas, NV
03 March 2021